The Space Trade Update

L. Paul Turner

- A Spacekind Series Book -

The Space Trade Update: Completely revised, updated, and amended Second Edition to The Space Trade, How to Develop Real Estate in Orbit, Make $Billion$ & $Trillion$, and Save the Earth, In the Current Age of Space.

Copyright 2018 by P. Turner Writer LLC All Rights Reserved

Cover design, courtesy of Michael Wrigley

Editorial Consultant, Adrienne Ellis

The following terms are considered trademarks of JP Aerospace: Ascender, Orbital Airship, Pongsat, Dark Sky Station, Orbital Ascender, Airship-to-Orbit, and MiniCube.

Any use of trademarks in this book is for educational purposes only, and is not intended to infringe, dilute, or harm.

Registered Trademarks used in this book, if any, are owned by their respective owners. Some terms used in this book are used in their ordinary meaning and not within their trademarked domains.

ISBN: 978-1-7237-1697-3

Table of Contents

Section A: Introduction

The Dream Comes First

Making Space Happen

Chapter 1: What's Stopping Us?

Section B: Spacecraft

Chapter 2: The Spinship

Chapter 3: Asteroid Mining Vessels

Chapter 4: Space Ferrycraft

Chapter 5: The Ascenders

Section C: Understanding Space

Chapter 6: How Objects Move in Space

Chapter 7: How the Space Economy Saves the Earth

Section D: Economics

Chapter 8: Investor's Step By Step Guide to Profitable Space Real Estate

Section E: Miscellaneous

Chapter 9: Space Development Timeline
Glossary
Addendum: Choice of Law and Society
CAVEAT ITAR
The Next Book in The Space Trade Series
About the Author
Dedication
Inspirations
Thank You
Final Note

THE SPACE TRADE UPDATE

Section A: Introduction

The Dream Comes First

We may all perish in some unforeseen global disaster, but it is better to look forward, and stand on the foundation of space adventure, courage, and wonder.

This book is for all who dream of seeing our species expand beyond our dear mother Earth, of saving our mother, and taking our place in the vast cosmos, that beckons us ever onward.

For many years, I struggled with the fact that no one knew how to make life in space profitable. Yes, low cost access to orbit was something everyone, including myself, believed had to come first. I finally stopped putting up with that false barrier. Indeed, I have written this book to explain how we can make a profitable economy in near space in the very near future, and without the restrictive purse strings of government aid.

I compared the advancements of engineers with the needs of investors everywhere. This book ties up loose ends of the dream, and takes us to a flowering of activity in orbit, and beyond.

Making Space Happen

To develop space real estate, you need to build spaceships. The biggest spaceship is a rotating shell called a spinship. Think of a giant bicycle wheel in orbit. The four spokes of the wheel are elevator shafts. You can see a simple diagram of this at the TheSpaceTrade.com.

Through the axle of the wheel, each expandable habitat enters the ship, to travel down one of the elevator spokes to the wheel's perimeter. Here, the habitats are positioned along the inside of the bicycle tire. A corridor in the wheel's metallic rim connects all habitats. This spinship is the community dwelling unit of the space trade. Inside it are industries that drive a productive community:

- One or more refineries that process ore from asteroids.
- Electrical distribution powered by solar cells.
- Food production by farms supplied with water, air, and soil from asteroid ore.

Like caravans of old, shipments of metals, medicine, and food, fly between all spinship stops and Earth markets. Imagine the traders of space who will work on these spaceships, like the sailors on oceans of yore. Here, people will dwell. Verily, this is orbital real estate.

We'll see, in stepwise fashion, how a profitable private sector economy is achieved in earth orbit and beyond. We'll understand how the space economy saves Earth from destruction. All this can happen shortly, within our current commercial age of space. Pioneer people, who take part in this adventure, are the founders of the space economy.

Chapter 1
What's Stopping Us?

"It's too costly to get to space." That's the mythos of today's aerospace industry. It's why we hang on to every little advancement in modern rocketry. It's the major reason we cheer on Jeff Bezos for space achievements, and Elon Musk with his SpaceX BFR (Big Fat Rocket). Everyone believes that myth. Here's what I believe: Space now. We have the technical know-how, and we have the prospect of amazing financial returns. This book explains how the two work together to achieve a sustaining space community.

Certainly, opening up space by a low cost method of gaining orbit would ignite a flurry of space development. That's one huge reason why I'm working with JP Aerospace, whom I believe is the conceptual leader in today's quest to design a low cost to orbit vehicle.

So, how do we construct the profitable, and sustainable, private sector space economy, and do it now?

The general answer is simple. Divide the project into its natural investment parts. Make sure each part is profitable. Throw in the vast riches of space, and you have the engine of unlimited solar system development.

Before we walk through the steps to a vibrant space age, let's examine what a mature space trade system looks like. This will give us a framework, by which to test all future economic progress.

Business on the High Seas of Space

Let's visit the year 2040. And, as we do, recall the English galleons, the Dutch fluyts, and Portuguese caravels of old. These ocean-going traders find their counterpart in today's space-traveling ferrycraft, that transfer goods and passengers from orbit to orbit. These ferrycraft are the sailing ships of space, and the spinships in orbit are their ports of call.

Spinships are the profit centers of the space trade. Let's take a tour inside a mature orbiting commercial campus.

Bert, an Earth-based business man with Spinship Ventures, contacts four of his colleagues and gives them each a ticket to a resort at the edge of space. The purpose is a four week work vacation. Tickets cost a mere $10,000 each, which is a steal in the year 2040.

Bert and colleagues gather at an ascender port, and Bert checks that everyone has arrived. There's Jake, Martha, Sally, and George. They board the winged airship, and hop into comfy seats. The pilot flies them up 140,000 feet. Here, the SV group deplanes to the dark sky station, nearly halfway to orbit.

After a few drinks at the sky lounge, the orbital airship arrives. Luggage is transferred, and our small group board the airship. As they retire to their cabin, they see the Earth curving away below them.

Five days later, arriving in low Earth orbit, the airship captain requires the group to enter the transfer capsule, and buckle in. She orders her crew to prepare for rendezvous.

The group watches as their window on the orbiting airship fills up with the fuselage of a space ferrycraft.

The group feels a jolt, and their capsule moves away from the long arm of the airship. Now floating in orbit, they accelerate toward an opening in the side of the ferrycraft. This is their berthing slot in the ferry's fuselage.

Once safely contained within the ferry, a burst of acceleration throws them back in their seats. As the hybrid engines move the ferry faster and faster, the now constant acceleration keeps them snugly seated.

The space ferrycraft begins a gentler acceleration, and the flight engineer opens the capsule airlock door, and invites the group to the lounge. Here, they mingle with the crew and are shown the path to the spinship in higher orbit.

Half a day later, they spot a large, bright hockey puck rotating in space. The spinship. It's a mile in diameter.

The flight engineer informs them the spinship rotation is only two degrees of arc per second. At that rate, the ship will take three minutes to make a full rotation. The slow spin provides a feeling of gravity for a comfortable and healthy stay onboard. This contrasts sharply with the pioneering, but freefall environment of the International Space Station, which had, of course, no centripetal gravity.

W-E-L-C-O-M-E. Lights on the disk welcome them. As the disk keeps turning, the letters rotate out of sight. The ferry captain heads his craft toward the hollow axle at the center of the disk. This is the entry portal to the disk's interior.

The captain asks the group to return to the capsule seating, and buckle in. The capsule moves out from the ferry, and into the hollow axle of the spinship disk. There it is held securely by clamps extending from the axle walls.

The group's seating compartment slips out of the capsule, and, after synching up to ship's rotation, comes to rest above an elevator shaft. The shaft is a radial that connects to the ship's giant rim.

The elevator comes up, and clunks to a stop. A door opens, and the group's seating compartment falls - two inches, to gently land on the floor of the elevator.

The elevator descends down the centripetal gravity well to the ship's rim, where the group feels nearly as heavy as they did on Earth.

As the elevator cab opens, they see a brightly lit corridor that angles upward out of sight. If the Mall of America were strung out along this common corridor, the group could not be more dazzled. Here, are retail shops, maintenance and service centers, space view restaurants, game parlors, and resort facilities. These support the ship dwellers in their business and domestic activities.

A trolley pulls up to take them to their hotel. Feeling like they've suddenly run a gravity marathon after days of little accelerative force, the group, on wobbly legs, exits the elevator, and takes their seats on the trolley. Off they go, toward the up-curving horizon of the common corridor.

Less than a minute later, the trolley stops at a door to their right. It's the Ceres Space Hotel. This is one of the many separate habitats that occupy the rim. They ride the elevator down two levels to the lobby. Here, the group feels heavier still. They check in, then split up to rest in their rooms before meeting for dinner at the Galaxy View Restaurant.

The next day, the Spinship Ventures divide into teams. This is, after all, a working vacation. Jake and Martha walk to the company's refinery habitat, and meet Henri, who is the foreman there.

Henri leads them downstairs to a lab of work tables and assay devices. He complains about the poor quality of the current asteroid mines. He suggests the mining ships prospect other asteroids that may be richer in hydrates, organics, platinum and gold.

Jake and Martha examine the raw ore as well as the refined metals. After comparing the ratio of ore mass to useful refinement, Martha is convinced that the foreman is right. A new asteroid must be found. They bid Henri goodbye. Their visit has been crucial to their continued mining success.

George and Sally walk down the stairs into a farm habitat that their company owns. The habitat is similar to a farm on Earth, with soil, vegetables, and watering system. It's leased to a family, and, as George and Sally arrive at the habitat level, they see a small field of green, and a cottage with a roof of grass. A goat munches on the roof. The Curbo family dwells here, surrounded by trees in a park-like setting.

The farm produces more than ten harvests a year. Planting a variety of crops, the farm family can feed 50 people from each acre's produce. Unlike Earth, where mother nature dictates soil type, here, soil conditions are made to specification, and the air is free of pollution and harmful pests.

The farm employs several students on two year exchange programs, providing work experience in space. In cooperation with other farms, the Curbos established a farmers market. It gathers twice a week in the common corridor to feed the community. Beyond the spinship markets, the Curbos sell a portion of their specialty produce to outlets in Paris and New York. The farm's excess oxygen is sold to other habitats, and denizens of the common corridor.

On a smaller concentric circle is a hospital. It is located closer to the center of the ship in order to provide half gravity. Bert, the group's chemist, is visiting his uncle in intensive care. His uncle awaits completion of his second printed heart. He likely should not go back to Earth, because, even with a new heart, the strain of full gravity may be too much stress for him. Contact with family members is encouraged, but remains infrequent. Bert's visit lessens his uncle's isolation.

After his family duty at the hospital, Bert decides to relax at the low gravity amusement park, where, with special wing gear and a gentle wind blowing, he can fly from point to point.

From the amusement park, Bert moves closer to the ship's axle. He is tasked with retrieving special compounds that are mixed at the no gravity level, on a floor occupying the center of the top of the ship. These compounds supply the company pharmacies back on Earth. He arrives at the unstaffed laboratory that is kept at zero spin. Bert takes the chemicals from the mixing device, and heads down the elevator to another lab to test the fineness of the compound. After confirming test results, Bert asks the lab manager to send the compounds to Earth via the next scheduled ferrycraft.

Three weeks of work and play later, the group boards another space ferry which makes rendezvous with an orbital airship, and returns to Earth.

This story is like an idyllic walk in the park-like real estate in orbit, but how is money made from the system, and how does it all come together?

Before we discuss how the financing works, we need to understand the five spacecraft types, and how they contribute to the space and earth economies.

Section B: Spacecraft
Chapter 2
The Spinship

The spinship is vital, because it permits the use of all resources from asteroid mining. Without the centralization of a spinship to process ore, asteroid resources risk disjointed delivery and processing. For example, if we are trying to process water from hydrates found in asteroid ore, and we're at the asteroid with an autonomous mining ship, a mechanical and chemical system must be devised that has no help from gravity. Some leaders in asteroid mining believe that by containing the ore in a crucible, then centrifuging the slurry to separate hydrates from the metallic oxides, refinement can be done at the asteroid.

After this process divides the valuable components, they might be shipped in different containers in a single ship headed back to earth orbit. On the way back, water is extracted from the hydrates, separated into hydrogen and oxygen, then pressurized and used to refuel satellites in earth orbit. Do you know how difficult the ore processing is? It's terribly complicated.

Precious metals extracted from the asteroid would be sent to earth by some means not readily available. Organics from the asteroid would be discarded at the asteroid after coming out of the centrifuge.

That disjointed system would require many rocket launches from earth to perfect the spacecraft necessary to produce the desired and efficient industrial processes in space.

However, with a place in space, from which to launch a thousand ships, we could centralize and control all the refining. This central place is the spinship. Once the ship is built, it becomes the factory for utilizing asteroid ore to make whatever industry needs for space operations.

The mining system could be improved by launching the experimental mining vessels from the spinship, instead of from earth. This would save time, and it would provide an industrial base in orbit, saving billions in heavy-weight launch costs.

The spinship does something else for asteroid mining. It permits mining and barge craft to focus on simply getting the ore and returning it to the spinship factory. This saves fuel, makes the mining system vessels less complicated, and therefore less expensive. It concentrates the ore processing where humans can intervene and apply the latest improvements in a single place, instead of having to modify every mining ship.

From the spinship, refined ore not used on the ship, such as some precious metals, can be sent to earth using a device made aboard the ship. A steerable drag system is one example. It's similar to an ascender orbital airship. We'll learn about that in chapter five.

How do we construct the first Spinship?

The initial construction of a spinship will require at least one launch from Earth of a centerpiece, to which the remainder of the ship can be attached.

Once the shell is complete, and rotation begins, launches by other companies will supply the expandable habitats to be fitted into the spinship's perimeter. Later, once an ore refinery is built inside a habitat, materials for expansion of the spinship shell can be supplied from asteroid ore, rather than from Earth. So among the first steps in space development is getting a supply route from the asteroid mines.

As we are about to see in the chapters on asteroid mining and space ferrycraft, mining and transport of asteroid resources is vital to both earth economy and to saving the earth.

Chapter 3
Asteroid Mining Vessels

In order to survive, the private sector space economy must provide first for all its inhabitants, then provide for earth dwellers. To do so, the space economy needs resources. Asteroids provide those resources.

What is asteroid mining, but the extraction of asteroid ore to obtain materials necessary for life and commerce. Extraction, not the moving of the asteroid, is the sensible way to mine.

The space booty from asteroid ore includes phosphorus, and precious metals, such as platinum, and gold. There is metal for spinship expansion, and minerals such as silicon, to be used in space, or sent to Earth markets. The most precious material asteroids can give us is water, most likely locked in chemical compounds. Water will be used in space farming, spaceship propulsion, and supplying oxygen for air, and hydrogen compounds for spaceship fuels. Asteroids offer carbon compounds known as organic material, essential to farming.

Why mine asteroids, when we have mines on the Earth? The reasons are legion. Earth mines are limited, tend to pollute, and are controlled by a limited number of nations and companies. We have only so much platinum, and it appears in concentrated quantities only in South Africa, and thus is subject to political control. We use an enormous amount of phosphates, which are vital to modern agriculture. In Florida, phosphorous mining leaves heavy concentrations of toxic and radioactive metals. Old open pit copper mines have left residues of heavy metals that result in deadly contamination of water.

The critical factors supporting asteroid mining are twofold. One is that asteroid ore can be returned to us by thousands of mechanically simple, autonomous vehicles. The second factor is that a large asteroid could hit the earth and destroy much of life, such as happened 65 million years ago.

Now, remember, not all asteroids near earth orbit have been discovered. NASA is working diligently to watch the solar system for inbound rocks. Recently, in 2018, asteroid 2010 WC9 came within 120,000 miles of the Earth's surface while clocking a speed of

29,000 mph. That flyby, well within the orbit of the Moon, may seem frighteningly close, but NASA says that there was no danger of collision. Asteroid miners, ranging far and wide, can act as early warning sentinels for comets and asteroids headed our way.

That 2018 whizzing rock was between 200 and 400 feet in diameter. About ten thousand years ago, a smaller meteor, estimated at 90 to 150 feet across, formed a crater a mile wide in the Arizona desert. Maybe you've heard of it. It's so iconic, they call it Meteor Crater.

The 2013 Chelyabinsk comet, came from behind the sun, and we had not enough time to deflect it away from earth. It exploded about 2000 feet in the air, causing injury to people on the ground, and windows in Chelyabinsk, Russia, to shatter. The upshot is this. We either help the human race by using the bounty of the heavens, or we suffer eventual destruction by space rocks that we did not mine.

Why not mine the moon?

The cost to get a miner to an asteroid and back is less than the cost of going to the moon and back. Yet, the moon is in our face in the night sky. We don't spot near earth asteroids with the naked eye. And, the moon holds more than visual wonder, or a place in human lore, it has minerals and water. A strong argument for mining the moon is that we can develop and improve rockets and other systems to both descend to the moon's surface, and to escape from the moon's gravity. We can learn to build safe and lasting shelters on Mars by building first on, or deep inside, the moon.

In fact, there is no reason to argue about mining the moon first, or the asteroids. Both mining efforts are proceeding at their own pace. That pace is determined by the exigencies of each operation, such as cost, distance, and whether the mission should be manned or carried out by autonomous vehicles. The private sector has chosen to mine asteroids first. However, asteroid efforts will soon be joined by moon mining, and possibly by people stationed on the moon. Recently, space directive number two was announced by the white house. It proposes we head back to the moon and establish commercial activities there and in space. And, if the government reneges, there are still major players, probably Elon Musk, and Jeff Bezos, who are intent on seeing humans permanently occupy space, including the moon.

Reality versus Prediction

Asteroid mining starts the space trade. This is not a prediction, it is a reality. Already, a prospector test satellite circles the Earth. This is the Arkyd-6, designed to look for minerals on our globe. Its progeny will perform the same examination of near Earth asteroids that lie ahead and behind us in orbit around the sun.

We would be well served by having a refinery in space to begin to utilize asteroid wealth. In fact, as previously stated, the least expensive way to complete building of a metallic spaceship in orbit is to use metal from nearby asteroids. This avoids excessive energy needed to boost building material up from the Earth. Of course, first we require a ship, smaller in design than the finished spinship, but large enough to house an ore refinery.

Advances in 3D printing of plastics and metals could permit the spinship to construct, in orbit, many of the refinery's components. And, because some ore processes rely on chemicals, rather than smelting, some refinery parts could use plastic, instead of metal.

When the first refinery in orbit is operational, precious metals could begin shipment to Earth markets by one of several methods, as we'll see in Chapter 6, discussing objects moving in space.

Here is a short list of reasons to mine asteroids.

1. Asteroids give us space development and sustenance at a cost lower than launching materials from Earth.

2. Asteroid mines are safer than earth mines, because mining is performed by autonomous mining ships, not people risking their lives in a mine. Asteroid mining operations could be monitored and directed from an earth base, prior to spinship completion.

3. Mining in space can and must be virtually free of pollution, unlike mining on and in the Earth.

4. Because it's a better alternative, asteroid mining may supplant Earth mining entirely.

5. Asteroid ore can be sent anywhere, to a spinship in low earth orbit, to a spinship in translunar orbit, or to Earth.

6. Mining in space provides an early warning system to prepare us against deadly comet and asteroid collisions with Earth.

7. Precious metals are more abundantly available in asteroids than they are on Earth, which contains most of its precious metal far beneath the crust.

What does a miner look like?

The first asteroid miners will be prospector vehicles made of ganged cubesats, such as the Arkyd-6. Each cubesat has the standard dimensions of 10 cm (about 4 inches) on a side. It's a cube!

Cubesats can be powered by small ion engines. They will go out ahead of Earth, in its orbit around the sun, and prospect for nearby candidate asteroids.

Along with prospectors are the mining vessels that grab onto an asteroid and drill for ore. These miners stay in and among the asteroids, moving to better ore candidates as they are discovered.

To service the miners, we have mining barges. These spacecraft constantly drop off empty ore boxes, and collect full ones from the miner. After taking all the full boxes it has attachments for, the barge returns to translunar orbit to drop off the full boxes, and pick up new empty ones. The barge hardly slows to accomplish these translunar maneuvers, then, it speeds on its way back to the miners.

Which asteroids must be mined first? And why?

The first asteroid to be mined could be one that occupies Earth's orbit, but lies 60 degrees ahead of the Earth. The angle is measured from the triangle that extends from Earth to sun to asteroid. That angle is about 60 degrees of arc. Behind the Earth, trailing in Earth's orbit are other asteroids, 60 degrees behind us. Together, these near earth objects are called the Apollo asteroids. They are the rock groups most easily reached. This means travel to them consumes less fuel than going to the moon and back.

In contrast, objects in the asteroid belt, which circle the sun beyond the orbit of Mars, are currently too far to mine. Better to improve the mining system closer to us, before venturing beyond. As you may have deduced, lowest fuel consumption is the main driver in deciding which asteroids to first examine and mine.

What if we don't find anything but iron and silicates in the near Earth asteroids? Well, we would be very surprised. There may be layers of primordial wet goo on any asteroid. The near Earth ones may have a sun dried crust on top of the hydrocarbon ooze. -

How will we get the rocks back to Earth?

The mature system has dedicated miners at each asteroid. These miners are serviced by space mining barges. These are vehicles that collect the full ore boxes, and leave the auto-miner with empty ore boxes. The barges form a conveyor belt heading back to translunar orbit to drop the ore into the bay of a space tug, and collect a fresh supply of empty ore boxes. The barges then head back to the auto-miners for more ore.

Why can't we just send humans to mine the asteroids?

We can send humans to mine asteroids, but the expense is great. To send a human being would require stores of air and food, equipment to process and recycle waste, and a living container for pressured atmosphere. The safety concerns are enormous. So, with today's computerized and specialized artificial intelligence, we can learn how to mine asteroids. Even so, many trips to an asteroid will have to made, along with constant improvement in mining vehicles, because despite the cover image on the first edition, "The Space Trade," we will need more than just shovels.

Who are the players in asteroid mining?

Wikipedia lists a number of companies, all of them private:

Planetary Resources (update: sold and collapsed)
Deep Space Industries (update: sold and collapsed)
Moon Express
Kleoas Space

TransAstra
Aten Engineering (going strong at print time)
OffWorld
SpaceFab.US
Asteroid Mining Corporation Ltd. UK

Chapter 4
The Space Trader Ferry Craft

The space trader ferry is a craft necessary to connect asteroid ore with spinship refin- ery systems, to connect earth and space markets, and to carry passengers from low earth orbit to spinships in higher orbits.

The first space trader ferrycraft is launched from Earth, to travel permanently and only in space. Later ferrycraft can be made in space by spinship industries. Moving between orbits, the space ferrycraft shuttles cargo, passengers, and asteroid ore from one point to another.

Passage Out of Earth

The space ferry saves fuel and cargo space on rockets launched from Earth. In order to travel from Earth to a habitat on a spinship in a low earth orbit of 300 miles (about 480 km), a passenger and cargo container goes through a series of travel stages. Going directly from Earth to a spinship orbit would consume too much fuel, and would overburden the rocket. The passenger and cargo load would be quite small. Even with a system of low cost access to space, going beyond a low orbit would be inefficient.

With the space ferry system, rockets coming up from Earth need not carry extra fuel to arrive at spinship orbit. These rockets can be met in low orbit by a space ferry. There, a standard pressurized container of passengers and cargo is transferred from the Earth-launched rocket and moves into the fuselage of the space ferry. From this point, the ferry travels into higher orbit to rendezvous with a spinship.

Arriving at the spinship, the passenger and cargo container is released from the ferry to travel under its own propulsion several hundred meters, where it enters the hollow axle of the spinship. Passengers and cargo exit the container to enter the elevators in the radial arms of the spinship. Meanwhile, passengers and cargo disembarking the spinship enter the container, which then returns to the waiting space ferry. The ferry's next stop may be the industrial spinship in translunar orbit, far above the moon's orbit, or the ferry's schedule may dictate a return to low orbit to deliver the new passengers and cargo to a rocket arriving from Earth.

What If There Are No Spinships?

If no spinships have been built, space ferrycraft will take ore from returning barges directly to low orbit, where the craft will deploy an ascender wing, or other sky float device to send the ore to Earth-based buyers. The ferrycraft may then use asteroid derived fuel received from the barge to refuel itself, and satellites in Earth orbits. However, the newer satellites are beginning to employ ion engines to keep them in orbit longer. They may require no refueling for years. A space ferry could help place satellites in proper orbit after receiving them from a rocket met in low earth orbit.

Meeting in Translunar

As stated in Chapter 3, the asteroid barge craft returning with ore, organics, and water from the asteroids will have to slow down to rendezvous with the space ferrycraft. The barge will expend the least time and energy if it can rendezvous without changing orbital speed. A ferrycraft in high translunar orbit provides the best match of velocity with the passing ore barge. The ferrycraft matches velocity with the barge, takes its load of asteroid ore and water, transfers empty ore boxes to the barge, then shuttles the ore and water back to a refiner and fabricator on a spinship in lower translunar orbit. Because the barge did not have to change its velocity much, or at all, it will efficiently return to the asteroid miner to collect more ore. In this manner, a translunar ferry provides rapid transit of ore to earth or spinship.

Ferry Craft Fuel Efficiency

Ferry craft are designed to be like switching yard locomotives. That is, their main function is to change tracks, or orbits. Because of the great energy needed to change orbits, the ferry will need big ion engines powered by the sun, as well as powerful chemical rocket engines, or a new engine that combines an ion engine with a traditional chemical engine. Even with this combination of engine types, transport time to and from low earth orbits may take days. Yet, ion engines alone are sufficient for hauling cargo.

Ion engine capability was demonstrated some years ago by a launch problem the US Air Force encountered. A rocket was part way to orbit, when the main propulsion engine failed to ignite. Yet, the satellite on board eventually made it to its planned orbit because monitors on the ground deftly utilized small ion engines on the satellite rocket, even though they were not designed as main thrusters. The ferry craft will have more powerful ion engines than did the USAF satellite, and thus arrive sooner at any destination.

If the ferry is carrying only raw materials, the long shuttle time might not be a problem, especially if enough craft are built and put into service to make for daily scheduled deliveries. If the ferry has passengers, the craft will use more of its hybrid chemical thrusters, in order to deliver passengers to a spinship before air, food and water provisions are depleted.

Chapter 5
Low Cost Access to Orbit

The Ascenders

We don't need low cost access to space, in order to profitably develop real estate in orbit, but it would certainly help. While we don't yet have low cost access to orbit, we do have aerospace companies, notably, SpaceX, JP Aerospace, Blue Origin, Rocket Lab, and many others, working hard and smart to lower launch costs.

Part of the high cost to access space is people, sending and keeping them safely in orbit. Because you don't have to feed and provide air to machinery, it makes sense to use autonomous fabricators to gather material from the asteroids, and to construct the wheel-shaped spinship.

The advent of very low cost transport to orbit will change how we construct stations in space. Earth launched materials could give us a number of small spinships, while asteroid material could be used to expand those ships.

Low cost to orbit will dramatically change how we live and work in habitats nesting inside the spinship shell. Factories in orbiting habitats will require special equipment and specialized and experienced technicians, construction personnel, and all sorts of service support typically associated with establishing an industrial economy. Materials for inflatable habitats are also among the items that must first come from existing earth-based industry. Yet, innovations in robotic fabricators, and using autonomous ships to recover ore from asteroids, will allow industry to grow aboard the spinship.

Refined metals and other material will be shipped to earth via ultra low cost drag devices. Workers will be able to travel to and from earth with greater frequency, as the cost of travel to and from orbit decreases.

We will have business competition, both in space, and on earth. The gates of heaven open wide, and millions of tourists rush to space for a one or two week vacation. Workers can stay for months, or simply a few days. This is relevant if it turns out that not everyone is just so gaw-gaw eager to live in space.

Living in space to work will be an opportunity for many people. The space economy will be community based, and that community will need technicians, farm workers, doctors, nurses, lawyers, religious leaders, hairdressers, grocers, and all occupations that earth communities generally have. New hires will be able to come and go because access costs are low. They can stay, because food costs will be low due to in-orbit farms, and because space housing costs will be low compared to space wages.

Space wages reflect the greater productivity of the average worker, who will have control over the machinery that does the work of many hands. The adjustment of living costs depends upon the natural evolution of trade between and among citizens of space, and their trade with earth-based markets. Costs will vary, depending on the usual economic forces of scarcity and necessity. Attempts at price controls by government, will no doubt be met with the usual black market economy. Avoiding such control by government, no matter how well-intentioned, will be vital to space market progress.

If low cost access to orbit means that humans will easily populate space in a constant stream, then life on earth will change rapidly, as earth-based companies compete for workers which the space trade economy demands.

What Drives Expansion of the Space Economy?

While one driver is the sale of asteroid material to earth-based industry, a second is the self-sufficiency of the mature spinship. Self-sufficiency can be achieved after the berthing of only several habitats. Self-sufficiency is a foundation for an expanding economy, because investment risks can be undertaken with the knowledge that the essentials for life are taken care of. Self-sufficiency does not mean living from hand to mouth. It means, that with a few habitats supporting life, the greater majority of habitats can be devoted to all other economic activities.

At the heart of that new marketplace will be those who invested early, and have garnered the experience and suppliers necessary to build and expand real estate in orbit.

We know that low cost to attain orbit fuels expansion of the space and earth economies. But, just how low will be that cost? In 2004, at the Space Access Conference in Phoenix, Arizona, John Powell of JP Aerospace predicted something like pennies per ton mile to orbit. That approximates historic Atlantic shipping rates. Exactly when will the price be that low? No one can pinpoint the exact time, yet it is coming sooner than most of us can imagine.

Historical statements about costs to orbit stand on the figure of $10,000 to get a couple of pounds to low earth orbit. Since the turn of the 21st century, the cost of getting to orbit has dropped. Notably, SpaceX has been able to reduce costs across the board, introducing innovations which previous cost-plus type operations had not had to consider. For example, the time required for loading fuel has been reduced. The transfer within the rocket of fuel to the engines has been sped up or made more efficient. The weight of the dry mass of the rocket has been reduced compared to its fuel weight. All these improvements could have been done in the last so many decades, but the financial incentive resided in getting more money from the government for a less advanced product. Government control of specifications for rocket design did not have the basic traditional American cost incentive to achieve the efficiency gained by, for example, SpaceX.

Times have changed. As recently as 2005, NASA estimated the cost to build a Falcon 9 at $4 billion. SpaceX built it for less than ten percent of that NASA estimate. This is encouraging news, and especially so because the Falcon 9 and its later sister rockets are not powered by some fantastic science fiction development, but stand on the trials and successes of rocket engineering from the last hundred years. In short, if Congress had given NASA the go-ahead for competitive commercial spaceflight several decades ago, we might have benefitted from more people like Elon Musk. Some in aerospace believe the cost to access space could be reduced to the range of $500 per pound, and even as low as $100 per pound, while still using rockets.

A Flood of Competitors

The truth about trading in space has been obscured, mainly by government programs that, by their very existence, took focus away from price, and misled a generation of space frontier hopefuls into thinking that costs were going to be insurmountably high for the foreseeable future.

Bulging from a procurement method like cost plus, the rocket industry moved at a snail's pace of new development. Launch to orbit costs remained high from the 1960's until the beginning of the 21st century. However, that is changing. Even so, a $50 million launch price tag is ridiculous when contrasted to Atlantic shipping rates.

Despite the easing of cost to orbit, government pork barrel still exists. NASA has been directed by Congress to provide a continuation of jobs in aerospace, which includes building huge rockets that take funds from other projects that many in NASA and the scientific community deem worthy. NASA is restrained from choosing launch companies on the basis of emphasizing price, because currently, Congress is still the body that directs NASA to decide who will build what, and what for. Even though SpaceX can use its new and improved launch system, the company was slated to get only about forty percent of the launch funds available for the Commercial Crew Development Round for trips to the ISS. Yet, this is probably a transitional situation, and one that benefits our need to keep knowledgeable engineers, beginning engineers, and a variety of companies in the space industry.

An example of a high cost project that some think will never get off the ground is the Space Launch System (SLS), part of the continuing effort by Congress and the administration to keep people from losing jobs. The SLS is a giant rocket which many in the aerospace industry believe should not be funded. It's a big mission to launch a giant payload. It is a stretching of our rocket engineering to an extreme. Perhaps we benefit from some extremes. While Congress and NASA are asking industry to do something that has never been done, namely, put a huge payload into space, the specifications for the SLS are relatively fixed. If the SLS project could only use the advances by SpaceX, perhaps through a licensing agreement, the SLS might get off the ground. Using the most advanced technology doesn't seem to be the intent of Congress in this instance.

The dream of the SLS has its roots in the early days of rockets. The famous space rocket leader Werner von Braun proposed the idea, and then later rejected it. Some will recall seeing an artist's conception in books of the 1950's, which showed a giant cone-shaped rocket filled with hydrazine, a rather nasty substance, but one that does offer relatively easy engine re-ignition in space.

While the SLS may be a pork barrel project, the rocket scientists working on that project are some of the best. Unfortunately they will have to wait to see their own designs and improvements expressed in some later commercial venture that leaves Congressional meddling behind.

As a matter of politics, we certainly don't need to ask the government to draw in all aerospace companies, all competitors, to a more costly system for launches. That's what happened in 1958, and through the Apollo moon program. We have seen that the federal government has, as yet, been unable to promote lowering of costs across the board. In fact, the lowering of costs will arrive on its own, from one or more causes. The first cause may be China on the moon, making lunar prospecting pay for itself. If this happens, and if the USA gets in gear, and uses competitive bids, low cost moon mining will be the order of the day.

The second factor for lowering costs occurs when competition for market share in the asteroid trade heats up. As space real estate becomes profitable in refining asteroid ore, and as fuel depots are built in high orbits, competition will lower the cost to launch, and the market will come to equilibrium mostly on the basis of reliability and market price. As the average cost to launch decreases and creates more space business, pork-barrel programs would only draw valuable workers away from the burgeoning private sector.

As we have seen in the news, alternative methods to reach orbit are being explored: the space elevator project; the Skylon hypersonic endeavor; JP Aerospace's concept; and others. However, if space real estate developers wait until access cost is very low, they will find that competition for knowledgeable workers and suppliers will restrict immediate profit potential. On the other hand, long term profit potential should remain high, even as space real estate reaches a mature stage when spinships venture out into translunar orbits, and orbits beyond it. The upshot is that those who wish to develop space real estate can stay ahead of the crowd, only by getting into the game early.

The Low Cost Method of JP Aerospace

What if you had a system of getting to orbit that cost as much as a boat ride across the Atlantic? What if the system was comparatively green as well? What if the accelerative g loads experienced by crew, cargo, and passengers were pleasant? Lowest cost, friendliest to earth, and most comfortable. That describes the launch-to-orbit design of JP Aerospace. The author admits he's been a fan of JP Aerospace since first hearing about it in 2004. That year, the author became an associate, and was invited to build flying systems that move us along the experimental path to low cost victory over gravity.

Using the complete JPA system, the general public will be heading up to orbit, because both the out of pocket cost and the environmental cost could be lower than a vacation trip to the nation's capital.

How to Achieve Low Cost to and from Orbit

Imagine the launch pad scene of a large traditional rocket, billowing polluting smoke, heavy with fuel, rife with risk to crew. Viewing the launch, a hopeful crowd cheers as the crew makes it safely to orbit. Now picture a winged aircraft flying efficiently to space like a rocket plane, yet with no giant billows of exhaust, and no stressful accelerative forces. This kind of aerodynamic flight to orbit is for comfort, safety, and price. It is also the clean air choice.

This airship-to-orbit system is the brain child of John Powell, president and owner of JP Aerospace. His book, "Floating to Space," describes the experimental path to floating and flying to orbit. How does it achieve orbit? A 3-stage patented method is employed, using ascender wings and a dark sky station.

An ascender, a V-shaped balloon wing filled with helium, flies crew and cargo from the ground to about 140,000 feet, where it docks with a dark sky station. The station is a large 5-fingered dirigible, digits spread like a starfish. It is buoyant, and maintains a high altitude range that varies with changes in atmospheric temperature. At the station, crew and cargo board a larger ascender, called the orbital airship, which flies to a higher altitude, perhaps some day to orbit.

The orbital airship utilizes specially designed hybrid engines, combining chemical and ion engine efficiency to achieve a new kind of propulsion.

Engines are embedded into the wings for maximum aerodynamic efficiency.

The orbital airship, by the nature of its size and shape, utilizes a low rate of acceleration. The chemical engines, which on rockets would expend their fuel in a matter of minutes, are modified to provide a lower thrust value over a longer period of time. To use less fuel than a conventional rocket engine, the ascender's engines include an ion component, which may derive energy from solar cells on the skin of the wings. The hybrids would achieve, over the course of several days, the required velocity to achieve orbit.

The acceleration load on the orbital ascender is limited to something far less than full earth gravity. For this reason, passengers will experience comfortable low g acceleration all the way to rendezvous with a space ferry. At the rendezvous, the airship's compartment filled with passenger and cargo is transferred to the ferry, even as the ferry moves its compartment to the airship, for a return to the same dark sky station, or another station on the other side of the world.

The passengers in the ferrycraft shuttle on to the spinship, where their compartment is transferred to an elevator that brings it down to corridors that link with habitats nestled in the perimeter of the circular spinship.

Section C: Understanding Space

Chapter 6
How Objects Move in Space

There is a common misperception that if a rocket is up high enough, say several hundred miles, there will be no gravity to pull it back down. This is akin to the captain in a science fiction spaceship calling for an all stop. In space, there is no all stop. This is because, if you are not moving in relation to the planet you are near, your acceleration will be in the down direction.

Even the moon experiences the pull of Earth gravity, and it's at an average distance of about 239,000 miles (about 384,000 km). That tug of Earth's gravity is what keeps the moon from sailing off to enter its own sun centric orbit.

You can imagine that at an altitude above the earth of a mere 250 miles (about 400 km), a space station would be not ten percent farther away from the center of the Earth than someone standing on the ground. In fact, the force of gravity diminishes with the square of the distance from the center of gravity, which, for our world, is the center of the Earth. This means, in Earth orbit 250 miles up, the force of gravity is still about 90 percent of what we experience here on the ground. If you weigh 100 lb. on the Earth, you would weigh 90 lb. at 250 miles up. So, why you might ask are people floating around the inside of the International Space Station. They obviously don't weigh anything. Where is gravity? Actually, those space citizens are in free-fall. They are constantly falling, along with their space station. Only the speed of their whizzing around the earth keeps them from immediately being pulled back down.

Yet, as we know, the ISS does slowly fall to earth, moving into smaller and smaller orbits. So, every so many months, the station fires up its engines and speeds back up to a higher orbit.

In the ISS orbit, where there is still 90% the force of surface gravity, that's quite a pull on a space station. The ISS ranges higher and lower in orbit, because of gravity, as we just mentioned, but also because of drag from earth's atmosphere. The old planned altitude of 220 miles (350 km) forced the station to expend about 19,000 lb of propellant per year to accelerate up to a higher altitude where there is less drag. The new planned altitude is 248 miles (400 km), which will require only 8,000 lb of propellant per year to keep the station in the higher orbit. That's quite a savings in fuel. The earlier planned lower altitude existed because the U.S. space shuttle could carry more cargo to the ISS, when the ISS was lower. With retirement of the shuttle, and with new rockets to reach the new planned altitude, the station will save fuel. At this new, higher altitude, the ISS will actually be going slower in its orbit, because, as previously stated, the force of gravity decreases with distance, requiring less speed to counter that diminished gravity.

Maybe you've noticed. The ISS must accelerate up to a higher orbit, but once there, the orbital velocity required to maintain orbit is less. This is because change of orbits requires energy. The higher the orbit, the less energy required to stay there. Hence, the station is now better off fuel-wise in the higher orbit.

Orbits: Location, Location, Location

Where to place the first spinship?

In order to prevent the need for a boost of velocity every six months, a spinship should locate in a high orbit. Initially, while under construction, the ship should reside quite low in earth orbit, to be easily visited by rockets loaded with construction material.

The final orbit should be about a thousand miles up, where less drag from stray atmospheric atoms exists. However, at one thousand miles up, the spinship would be deep within the Van Allen belt. Here, it would be subject to intense radiation. The belt starts around 500 km, or about 310 miles. At about 58,000 km (about 36,000 miles) you are outside the belt, but would be experiencing radiation of a highly destructive nature from cosmic sources. What to do?

The first spinship should have its permanent orbit where ferrycraft can reach it easily, and where drag is as low as practical. That regime places the ship slightly higher than, but close to, the orbit of the ISS. Spinship heights would range between 470 to just under 500 km (about 292 to 310 miles).

So the initial spinship should be placed in low orbit. However, this presents another problem. Mining barges returning from the asteroids would have to greatly slow down to deliver material to the ship. This uses a lot of fuel. Even so, at first, until we establish an efficient refining system onboard the spinship, the 500 km orbit is the best scenario.

As the first spinship refinery is completed, space ferrycraft would travel to translunar space to rendezvous with barges speeding around the mining circuit. This places the burden of energy use on the ferrycraft, instead of on every returning barge.

Later, a spinship, hardened against cosmic and solar wind radiation, could be placed in translunar orbit, making efficient use of fuel by the barges returning ore from the asteroids. Space ferrycraft would shuttle between translunar and the spinship in low earth orbit.

Where to Put Subsequent Spinships

Some have suggested L5 as an ideal place to orbit a space station, or spinship. L5 is LaGrange Point 5, a high orbit beyond geosynchronous, tucked just under the moon's orbit, and trailing the moon. Dr. Gerard K. O'Neill thought that his Island One space colony would be in an orbit somewhere between the Earth and the moon, beyond geo-synchronous, and in an orbit lower than L5. It turns out that at the L5 point nearer the orbit of the moon, there is gravitational influence from the sun, moon and Earth. This is not the stable point that popular opinion believes it is. Locating in a high orbit inside of the L5 point would give an advantage to ferry craft returning from mining barges met in translunar space. The ferry would expend less time and energy in changing speed to match the orbit of the spinship. Yet, a disadvantage of a high orbit outside a Van Allen Belt is the need for more shielding to protect inhabitants from cosmic rays and sun flares.

Whether the spinship orbits just below L5, or much lower at 310 miles (500 km), a ferry craft system is needed. This ferry system is the least expensive way to move asteroid material to any spinship, and to Earth markets. Ferrycraft can shuttle ore under autonomous operation, avoiding the cost of added weight of staff, fuel, water, air, and food.

Orbital Perturbations

Choosing an orbit for a spinship? What about gravitational perturbations from larger bodies such as the Earth, moon and sun. This factor is especially important for a rotating body such as the spinship, because rotational balance is influenced by gravitational changes. The exact magnitude of this influence is unknown, but depends upon the density, size and resilience of the spinship design, construction, and controls for maintaining the ship's flatness of rotation.

Another concern in orbit is related to factors present with any spinning body. Because the spinship is rotating, albeit with slow speed, it's vital to keep the ship balanced to avoid wobble. Disturbance from gravity may be small compared to anomalies in the energy of rotation. The problem is that rotational energy in one plane tends, over time, to be distributed to all planes. A top wobbling erratically as it slows down is an example of this phenomenon. To minimize this problem, many have suggested that a counter rotating mass be employed to stabilize the spinship. This is a worthy suggestion, even though designed rotation is so slow, that it takes three to six minutes to make a complete turn.

Designed time of rotation depends upon diameter size, material strength, and sufficient gravity feel that supports people, in their laboratory, in their production work, and in their health and recreation time.

The slow rotation of the ship is designed to eliminate any long-term Coriolis Effect on the inner ear. Dizziness from rotation would be generally avoided.

While a community of habitats, located in the large perimeter of a spinship, may be better served residing in low earth orbit, later ore refinery ships in high orbit near L-5, or in translunar orbit, could have a smaller disk diameter to save on construction costs, and could have a lower gravity feel than the one earth gravity feel on the community spinship.

The refinery ships could have special transient quarters, essentially small habitats, constructed at an extended diameter, providing a higher gravity feel for sleeping. The refinery crews could receive regular supplies from a full diameter spinship in low earth orbit. Scheduled ferries, which transport product from the refinery ship, would also carry workers back to lower earth orbit for time off.

Delivery to Earth

Asteroid ore can be sent back to earth via several means. The methods, that we are about to look at critically, were inspired by one of several talks given by Chris Lewicki of Planetary Resources.

1. Traditional Re-entry Vehicle: Apollo Style

Like the Apollo missions re-entry method, developed in the 1960s, an ore delivery vehicle would use a heat shield with an ablative surface. The nose cone is sloughed off by heat and air pressure during the fall through the thicker and thicker atmosphere. The heat shield is attached to the most forward facing point of the ore container. To be truly cost effective, this method requires manufacture of heat shields in orbit.

The nose and underbelly of an ore barge could be tiled with a material highly resistant to heat and air pressure. The tiles might be similar to the former space shuttle tiles, although changes in shield material have advanced the efficacy of this traditional re-entry method.

This system of re-entry successfully delivered Apollo astronauts to a targeted area at sea. The system was used by the Russians, except that they landed on the ground. Both American and Russian descents used parachutes in the lower atmosphere to reduce descent speed.

2. Re-entry by Retro Engine

A retro engine descent uses thrust in a non-ballistic return vehicle. Here, the cargo container is connected to a thrusting engine that slows the container to a speed that prevents burn-up and dispersal of contents. Basically, this is a vehicle that has a low ballistic coefficient of entry into the Earth's atmosphere. This is another way of saying that if the container is slowed prior to entering the Earth's atmosphere, and continues to be slowed by running an engine during descent, the material can be brought to Earth safely in one piece. This kind of entry would be aided by a large surface area to take advantage of air resistance.

Typically, in a tail first descent, the vehicle slows to a speed where heat and compressive forces have minimal effect on the skin of the spacecraft. The descent may take several orbits around the Earth. At each orbit, the craft is closer and closer to the Earth. At some point in the high atmosphere, the craft fires its engines to slow it to maneuvering speed in order to steer to a safe landing on a flat desert or at sea. This system offers maneuvering capability to land precisely on land or at sea.

3. Inflatable Drag Devices

This concept requires inflation of a sphere, or large area balloon that would slow the container in its orbital descent. Basically, this is a device relying totally on drag to slow the container to a safe speed. Ore attached to a sky bag, or skyfloat, would be subject to the whims of re-entry point, and to speed and direction of release. This is similar to unfurling a parachute. However, a parachute without sufficient air density will not unfurl. Waiting for the vehicle to get to thicker atmosphere before deploying a parachute risks ablation and fire. High speed deployment of a parachute at altitudes low enough to be dense with air, would rip the parachute apart. Luckily, the air drag system works for vehicles that are inflated large enough to gather sufficient drag at high altitudes.

A more intelligent descent is by ascender wing. The ascender wing is an inflatable drag device with the ability to steer. It's basically a large balloon in the shape of a swept-back wing, to which is attached a cargo container. The balloon wing is inflated with gas from organic compounds obtained from asteroid ore. The wing descends by making tighter and tighter orbits, using its size and aerodynamic shape to produce drag to slow its speed. It essentially bumps into the atmosphere over a longer period of time than the traditional heat shield method. The wing is like the ascender described in Chapter 5.

Use of ascender wing allows for precise navigation, such that a specified destination almost anywhere on the planet could be reached. The descent would be enhanced by using a gas, such as helium, or possibly methane, which would give the wing buoyancy. Use of hydrogen gas would become problematic when lower altitudes were reached, because hydrogen and atmospheric oxygen can easily combust.

Even though buoyancy may be desirable at lower altitudes for maneuvering, the usefulness of gas at high atmospheric entry is due simply to the maintenance of inflated wing size and shape. Even at very low pressures, for example, one pound pressure per square inch, the drag shape could be achieved. This means that very little gas in the inflated wing can effect precision deliver of ore to earth.

4. Space-built Cone Decelerator

This device is simply a cone shape made from asteroid mined material. The cone is the heat and ablative shield and the drag mechanism combined. Several shapes similar to a cone could be employed, that would allow steering of the re-entry container. A navigation system could ride within the cone for a controlled low ballistic entry to the atmosphere. Given the proper material, the navigation system could, at proper altitudes and velocity, alter the shape of control surfaces to direct descent to a specified region on land or sea.

5. Foamed Metal

A foam sphere of platinum metal several meters or more in diameter could make the jump from orbit to the ocean without breaking up in the atmosphere. The manufacture of such a foam ball would take place in space. How that might be accomplished is a mystery. If the ball could be shaped to provide control surfaces, navigation control could steer the ball more precisely than a simple drop from space. With a calculated drop from orbit, assuming no control surfaces for steering, the ball might be made to land in a rather general area.

Chapter 7
Save Earth

How the Space Economy Saves Earth from Pollution, the Greenhouse Effect, Asteroids and Comets

> Some people, we might call them earth-only environmentalists, insist that we perfect life on earth before gallivanting around the solar system in spaceships. What answers can we give them?

1. Earth life will never be perfect.

In fact, life is getting worse. Well, kind of. We on earth are moving in both directions at the same time. On the one hand, we're moving toward greater use of all fossil fuels, leading to increased atmospheric methane and carbon dioxide. On the other hand, we are finding ways to use carbon dioxide. Some companies make fuel by using the ingredients in carbon dioxide and natural gas. We are finding ways to absorb carbon dioxide in the making of concrete. We are devising solar arrays that will compete with oil-based fuel. We are, in short, making progress on the environmental front.

The political front is different, as usual. While we seem to have eliminated most of the fear of nuclear attack from the former Soviet state, who knows what might be thrown our way by a smaller country, such as North Korea, or a bigger nation such as Iran, or goodness forbid, China.

The point is, we would do well to improve life everywhere we go, even into space.

2. We can save the earth from pollution and greenhouse gases, now. We simply move polluting industry to space.

This just doesn't sound feasible. We can't move existing factories to earth orbit. The amount of pollution from rocket exhaust would conceivably increase pollution at a rate unheard of, even in modern times. Lifting each part of each factory would take a great deal of time. Besides, the factories would have to be supplied by earth materials, and that would mean using more rocket fuel and spending more energy to produce the same goods that were produced on earth.

Of course, while the proposal above is ridiculous, factories will at some point be built in space. They will be built from materials already in space, such as asteroid metals, and will use energy from the sun and from asteroid organics. Will they replace earth factories? Space factories may deliver some product to earth, but by and large, earth-based factories will continue to exist. However, their rate of growth should slow down.

3. Space development will blow up natural bodies whizzing toward us from deep space.

Here is the movie version of saving the earth. A giant asteroid is found to be headed toward earth. A spaceship is dispatched to blow it apart.

There are several things wrong with this scenario. How did we find out the asteroid existed and was headed our way? Maybe we had the best equipment in the best of all places, in orbit around the earth, the moon, the sun, and around the main planet known to attract asteroids and comets, namely Jupiter. And, would we want to blow apart the asteroid? I don't think so, because that could separate large pieces of the asteroid so that many would hit the earth. If the asteroid or comet pieces were quite large, we might destroy a number of favorite earth places, like a city where we live. It would be cleaner to send an existing barge craft to intercept the asteroid, keep the asteroid intact, and pull it away from the earth, and maybe even mine it. Better yet, we could target asteroids known to be future threats, and mine them well before they hit the Earth.

4. We are unable to predict all the consequences of permanent space habitation.

This is a correct statement. The concept applies to every new endeavor, and especially to space endeavors. The consequences of space entry have been both beneficial and detrimental. Weather satellites, beneficial. Rocket launches, polluting but useful. Yet the benefit side has big benefits to it, such as protecting the earth from sun flares, errant asteroids and comets. Another benefit is reducing Earth-based pollution. In short, permanent occupation of space is a very good thing, and that includes spinoffs from space development, which will, no doubt, continue.

Of course, there are risks with any space endeavor. Risks reduce when no ballistic rockets are needed to get us orbit. On the other hand, rocketry is still improving.

Someday a rocket flight may be as safe as a typical commercial airline flight. And, someday soon, friendlier forms of space ascent may become the norm.

5. What about objects falling from orbit? Don't we have enough debris up there already?

Throughout the decades, debris has been left in orbit. Nuts and bolts, exploded satellites and various other objects pose a great hazard to people and satellites passing through those debris orbits. A great deal of effort has been made to track each particle of matter that could cause damage to an existing satellite. Every launch is planned so rockets can avoid debris. Planning is vital, but wouldn't space be safer if we cleaned up the debris?

I have to applaud the European efforts to plan clean up of space junk. Clean up, rather than dodging, is the best way to protect all users of space. Of course, new debris appears from time to time. Sweeping the dangerous cobwebs from the sky must be a constant endeavor, especially as the number of flights to orbit increase in the years ahead.

So debris bits in orbit pose a hazard to those in orbit. What about us down here on the ground? Some objects burn up before hitting the ground. Others explode in the atmosphere, like the 1908 Tunguska event, and the Chelyabinsk meteor of 2013. If an object in space has enough speed and density, like a space rock, or metallic debris from a satellite, it may delay great slowing in the atmosphere, and continue to keep speeds that heat it up and burn it up before it hits the ground. This assumes the object is not so large and slow that only a part of it burns up in the atmosphere, leaving the remainder to crash into the earth, like what happened to Skylab, which left parts of itself strewn across Australia in 1979.

6. We are afraid of doing things that are out of the ordinary.

Some people simply don't want to think about outer space development. They are afraid of it. They are confused by it. They don't trust others to take care of it without ruining things here on earth. This is probably the most widespread and powerful reason why people don't want anyone to go to space and do great things. Those who must go forward to space do so with courage, or not at all. Every frontier appears to have this problem of fear. Perhaps the fear is simply a normal reaction to the unknown.

7. Saving the earth.

There are a number of ways that space development saves the earth.

Steering Comets and Asteroids

We will indeed be tasked with shoving aside asteroids, or chipping away at them by mining, to prevent them from hitting earth and destroying all life. In order to deal with asteroids in a timely manner, we will need resources in space. Without a robust space industry in orbit, we would be unable to deal with all asteroids. At the current time, some high speed asteroids, and comets from afar, may be headed our way. These fast-moving space chunks are often impossible to detect in time to push them away. However, a series of early detection, quick action spacecraft beyond the orbit of Jupiter or Saturn would provide the early warning and action required to save the earth from bombardment. Even asteroid miners, ahead of us in Earth's orbit around the sun, can help spot rocks headed our way from other parts of the solar system.

New Devices and Other Spinoffs

The development of space will require new, autonomously operating machines to do all sorts of things for us, from building habitat enclosures, to operating complete recycling systems, to repairing and maintaining solar electrical arrays, to making farming more productive. New devices and machines will be invented because there is a pressing need for them in outer space. Thus, space-developed machinery may exceed that which might be developed on earth. Inventive leaps, making systems that go beyond the next step needed on earth, propels earth's economy, and improves life for everyone.

Improved Traditional and Inventive Agriculture

Earth needs better farming. Space can provide farms with the ability to protect one farm from contaminating another. Farms in space can provide a helpful mix of non-GMO, USDA organic, traditional

California organic, and soil and fungi specific environments for growing specialty foods that are yet to be developed. In space, you can design soils, or use special humus grown from particular earth soil fungi. You can have abundant multiple crops from the same acreage month after month, producing more food per acre than the best farms on earth. You can air-drop emergency food supplies from orbit to areas on earth that might be experiencing drought, flood, hurricane, earthquake, or other natural disaster. You can grow bigger produce in low gravity habitats, or develop flower and other plant varietals that cannot be grown on earth.

Pollution Reduction

Earth needs help, especially in lowering levels of mercury from natural coal mine fires, and levels of methane, and carbon dioxide. Help from asteroid mining could reduce the amount of digging in the earth for some rare and important metals. The need to refine metals on earth would be lessened, because space derived metals would be refined in orbit, entirely avoiding pollution.

Power the Individual with the All Recycling House

Pollution can be reduced with the development of the all recycling house. Yet, the all recycling, or self recycling house can do so much more. Such a house on earth would generate its own methane fuel from food waste, use its own electrical energy from solar and wind sources, and derive heating and cooling from temperature differences between underground and the air. In short, the all recycling house is the self-sufficient house.

The farmhouse of yesteryear is the grandsire of the self recycling house of the near future. The farmhouse did not recycle its air and water; the all recycling house will. With an all recycling house, anyone can live off the grid. If everyone had such a house, we would leave no footprints, and there would be no grid.

This new house is absolutely vital to long term occupation of space. It is the self-sustaining space habitat. If we can develop that habitat, we will have saved much trouble for municipal services. Garbage collection, sewage treatment, water supply, and centralized energy systems would be unneeded. We will have avoided the construction and expense of new centralized grids. We will have softened the effects of business cycles on the average worker by making us all more self-sufficient. Today, we have many of the essential technologies for the self-recycling house. We just need to put them all together.

Earth Watch, Earth Talk

The earth is constantly in need of more satellites for communication, weather observation, and geo-positioning (GPS). We need more sophisticated ocean and farmland monitoring, volcanic and tectonic activity detection, and nuclear and other missile launch reporting. With the advent of a permanent station in space, and space mining, these satellite devices could be built in orbit, reducing the need to launch costly polluting rockets from earth.

Sun Block

The earth needs protection from the solar wind, and the sun's violent flares and magnetic storms. Especially important is protection from coronal mass ejections. These are masses of material thrown out from the sun. Some of them hit the earth's atmosphere and cause spectacular displays of color in the aurora borealis. The ejections occur with great frequency every century, especially during the solar maximum portion of the traditional eleven year sunspot cycle, when sunspot activity is highest. Less violent but more prevalent are solar wind storms that can disrupt modern internet, radio, television, and communication lines, broadcasts, and microwave relays, and even corrode pipelines in the middle of nowhere. A station in space, stable, with artificial gravity, and protection from the sun's flares, and from the solar wind, could be instrumental in protecting earth satellites from solar disruptive events. Backup satellites on the spinship could be retracted under the ship's shielding during an attack by the sun's increased activity.

Some early warning satellites are already in place, watching the sun for unusual activity.

There are a number of recent violent sun storms that have caused disruptions. If our vast grid of power, internet, communication satellites, and electrical power were to suffer such onslaught, like the Carrington event of 1859, we would see major destruction costing trillions of dollars. Spinships could form a backup web to keep business running, as the earth systems were repaired.

Sun Block Protection from Solar Flares

Protection from solar events, as well as from overheating due to carbon dioxide, or from the sun's varying output, could come from adjustable shields in place around the northern polar region. The shields could adjust to let in some or all ordinary sunlight, but prevent an excess of solar particle activity from reaching the earth. The south polar region, when affected by sun flare or magnetic storm events, does not normally affect as many people, nor is the effect as severe because dense populations are far from the pole in the southern hemisphere.

Yet, placing shields around any part of earth is controversial, even among space scientists. One might imagine a shield of thin metal, like a mirror backing, collecting the sun's charged particles before they descend to earth, where they might burn out electrical power lines, topple telephone poles, destroy radio, television and internet equipment, or possibly shock people near electrical wires, even as the Carrington event delivered electric shocks to many telegraph operators. However, there is no guarantee that such a shield system would protect the earth, and the experiment could prove as dangerous as the natural phenomenon it was designed to protect against.

On the other hand, shields come in various sizes and shapes. A temporary shield of sulfur dioxide sprayed into the high atmosphere could protect the north polar ice from excessive melting, much like plumes of sulfurous compounds from volcanoes cool the earth. However, this proposal also has its critics. Indeed, aside from the probability of atmospheric pollution, the sulfur shield might offer no protection from a stormy solar wind or solar flare event.

A simpler method to protect us from sun flare storms would be to use an early warning system. This system would give us enough time to employ widespread measures such as shutting off electrical grids for the duration of the storm. While present sun observing satellites do a great job, permanent stations in space could support an entire system of space sentries near the Apollo asteroids to give the earliest warning possible.

The best protection, aside from space-based solutions, or the advent of very local power, would be to send underground all our electrical transmission lines. To prevent lengthy transmission of energy bursts caused by the sun's particle storm, the number of turn-off switches in the grid could be increased.

8. Earthlings in the Void: Earth Renewed

All the above steps for saving the earth would be unnecessary if we had an effective return to a natural, park-like earth. This result is possible when we have the choice to live on Earth, or off Earth. Why? Population control would be unnecessary. With so many choices of where to go and what to see, many people would choose to explore the solar system. With earth as our ancestral mother, all humans would have an interest in her good health. The Earth could become free of all unsightly buildings, contraptions, power lines, pollution, and crowding. Yet, let's not paint too rosy a picture. The expansion of industry in space does mean that more jobs will eventually open up in orbit than on the Earth. People needing work will have it, but they may need to ride to space to get it.

9. Space Pollution.
Solar Wind: The Great Recycler

What about polluting space? Are we going to have a junkyard solar system? Recall that all material in space has value, even polluting chemicals, because with unlimited sun energy they can be transformed by devices aboard the spinship. In the unlikely event that something does escape from a spacecraft, whether rock, or dust, it must be cleaned up. Space cleanup ships which patrol the orbits would be dispatched to the polluting area. If the pollution, in the form of gas, escapes from a ship, the gas will be dispersed by the natural action of gases, much as are the tons of hydrogen and helium that escape from the Earth into space every day.

Some might think that the solar wind, that breeze from the sun of high energy particles and radiation, might transform any escaping pollution. In time it may, but only if the pollution were to feel the full brunt of the solar wind. The earth's magnetosphere protects our atmosphere from being whisked away bit by bit by the solar wind. This suggests that any pollution escaping from a spinship might stay around to be swept up by cleanup ships, instead of being taken away to be recycled by the solar wind. In any case, spinships would be designed to contain and transform pollution, not to release it. Pollution is another reason why spinships in the long run must prove to be the ultimate recyclers of just about everything. Applying that recycling know-how on the Earth gives us a cleaner place to live.

Section D: Economics

Chapter 8
Investor's Step By Step Guide to Profitable Space Real Estate

The Big Four Active Engagers

Engineers, Investors, Brokers, and Renters are the essential people that make profitable, the private sector space trade. Presently, most of our economic eggs are in two baskets. One basket is NASA and the US Air Force. These governmental entities are doing something in space. The other basket is all other private customers who want satellites in orbit. These customers are mostly in the communication and data fields.

What can we do now to broaden the private sector demand in the space economy?

Today's technical know-how allows construction of a spinship. This rotating shell in space, populated by relevant industries, can be the profit center that connects asteroid mining with Earth-based customers. With a single spinship in orbit around the Earth, the private sector space economy would usher in a new age of wealth and commerce for all of the world's population.

Financing construction of the spinship permits early investors, renters, and brokers to own, operate, and profit in orbit. We're about to learn how.

The Major Financial Players

The Spinship Investor-Owner
The Intermediary Investor, or, Broker
Tenants: The Habitat Renter, or User.

The Spinship Investor-Owner

The spinship comes into being when the first-tier principal investor overcomes three major hurdles to space development.

First Major Hurdle: Construction Plans and Price

Before anything can be built, a set of engineered construction plans must be developed by competent space engineers. Then, estimates and bids to construct the spinship in orbit can be invited from the best qualified companies.

Key to spinship design is the concept that only the shell need be built. The habitats that occupy the shell are built, flown up, and modified by others, namely, renters aided by brokers.

Second Major Hurdle: Find Tenants

Once drawings are complete, bids for construction are called for. This establishes the price. The next step is to find renters who will occupy habitats on the spinship at a price they can afford, and at a sufficient rate of return for investors.

The first renters to sign up must be ones that provide the essentials to life on the spinship. The essential renters will launch the first spinship habitats into low earth orbit. The group of essential renters must provide: Electrical power, refinement of asteroid ore, food, water, air, air pressure.

The first tenants will have a symbiotic relationship to each other. The electrical tenant will use solar cells on the surface of the spinship to supply the first essential renters with electrical power.

The agricultural renters supply other tenants with food, water, and air from farm habitats relying on asteroid resources. Yet, at first, necessaries will be simply hauled up from Earth.

As asteroid mining begins to deliver ore to the spinship, the refinery tenant will process that ore to obtain organic material, water-bound compounds, and oxygen. This tenant becomes the supplier of soil and water used by the farm habitats to produce food, air, and clean drinking water for all habitats.

Once these essential tenants are signed up, the spinship can accept other tenants, including those providing service, maintenance, lodging, and retail sales.

Most tenants will contract with the onboard farmers to supply food, water, air, and consequently, air pressure. For example, a hotel tenant would want to concentrate on providing living quarters, and leave the business of food, water, and air supply to the farmer.

Other renter-users form a long list, and include:

Rocket fuel suppliers. Research companies. Makers of spaceship parts. Makers of household products. Service technicians. Wholesale and retail suppliers. Consulates of foreign governments. Medical personnel and facilities. Hotel and resort lodging for space tourists and workers.

With the cost estimate in hand for building the spinship, a schedule of rents is calculated. Then, before a renter can secure a place on the spinship, the renter must agree to pay rent, pay a deposit of good faith, and show, within a reasonable time, that they can secure a space habitat, and can pay for its launch to the spinship.

The typical tenant will connect with a space habitat supplier, and a designer of interiors, in order to secure the necessary build out of tenant specific equipment.

Finally, the rental contract states the priority of placement. Essential tenants, as stated, are accepted prior to non-essential tenants. Tenants on the waiting list will be slotted into spinship habitats when gaps in service providers, or space available may dictate.

The Third Major Hurdle: Funding the Spinship Project

Remember, we said there are three major investor groups necessary to fund the building of the spinship and its habitats. These are the major investors, the essential renters, and the intermediary investor-broker.

The first-tier major investors may be a small group, perhaps as few as five or ten wealthy individuals. They will be the owner-operators of the first spinship. They will at some point probably form a company that can float a public offering of stock shares. However, initially, these major investors are on their own. They may wish to keep it that way, depending upon the prospective profits projected over time. In any case, these first investors can bring in ordinary, that is, non-wealthy, investors through a system similar to a Kickstarter or an Indiegogo project, where these additional investors can own a piece of the equity in the spinship. Yet smart investors may want to avoid SEC rules entirely, by simply keeping ownership to themselves. If the first spinship succeeds in making a good return on investment, the stage is set for further spinships. Yet, all spinship investment can succeed only by satisfying four factors vital to investors.

Four Investment Factors Vital to Spinship Development

1. Liquidity

The most important factor in any investment is liquidity. Can I get my money back by selling my interest to another entity in the broader marketplace? If the answer is yes, you have liquidity. If the answer is no, don't invest in the first place.

Liquidity, for an owner in the development of the first spinship, comes when other investors are clamoring to get an ownership interest. Liquidity opportunities happen at events in the stages of progress toward completion of the spinship shell. There are three major early times when this may happen.

First Liquidation Opportunity: when the buzz hits. Here is the time of early adopters. These may be about 16 percent of the total group of eventual owner-investors. The first buzz hits at the time of announcement of the project, and at each addition to the group of charter investor-owners.

Second Liquidation Opportunity: when plans and bids begin, and when they are completed. The time of completion of taking bids may be extended, if the group of investor-owners determines that, 1. Other constructors have developed superior construction methods that lower the cost of construction, or, 2. One or more governments seek tenancy. This chance to liquidate exists because some investors on the sidelines were waiting for the funding security provided by a long term lease to a governmental agency.

Third Liquidation Opportunity: when tenants start signing up. Tenants are the entities who pay back the investor-owners through the steady flow of rents charged for occupying and using habitat space aboard the spinship shell. As the number of tenant entities seeking space increases, the certainty of funding completion of the spinship shell increases, and non-early adopter investors begin to seek ownership interest in the spinship shell.

Fourth Liquidation Opportunity: as more and more tenants sign up, the deposit amount in escrow increases. At some point, the deposit amount, which is not refundable to the would-be tenants, will exceed the amount which the owner-investors have paid into the development fund. This point can be reached by any number of mechanisms. For example, the competition among tenants for space on the first ship, may force tenants to bid up the price of rent, exceeding the total rental amount originally envisioned.

Fifth Liquidation Opportunity: the fifth opportunity to liquidate arrives as third party brokers enter the mix. Early brokers arrive well before tenants may show up. As word spreads that tenants are being signed up by brokers, the late arriving brokers will push the competition for habitat space to new levels. An early investor-owner of the spinship may see a profit at this stage of liquidation opportunity.

Sixth Liquidation Opportunity: the sixth opportunity to liquidate comes as the spinship shell is being constructed and all is going well.

2. ROI Certainty

The second vital factor to spinship development is the certainty of return on investment. How certain is it that the first spinship will succeed? It's never been done before.

3. Speed of ROI

The third factor to spinship investment success is the shortness of time period in which the investment returns its cost. If the first project takes ten years to get a return on investment, that is too long to wait. Investor confidence would falter, and several investors would fall out, and possibly be replaced, or not.

The speed of ROI is determined by several factors, most prominent of which is the schedule of construction. A special element here is that the first part of a large spinship can be built with a smaller diameter. Renters could occupy the smaller uncompleted ship, and pay rent, even as construction continues to expand the number of berths that habitats rest in.

4. Continuing ROI

The fourth factor is that of proving the continuity of rents that constitute an unbroken stream of profits above and beyond the original investment. Once the ship is operational, continuation of rents depends upon market life on the ship. That market life will determine which tenants sell out, and which stay. In short, no single individual need determine or plan the market. The market speaks for itself, and through time and change, finds the mix of renter types that provide the synergy for market sustainability.

To sum up ROI, investors would want a quick, certain, and continuing return on their investment. The most important factor at the start is the certainty factor. Good planning, and experienced entities, provide confidence. After all, each new construction project, whether on earth or in space, brings together, often for the first time, experts in every field.

Funding the Renters

Because tenants have the responsibility to launch their habitats, and have them placed into the spinship, they bear the great costs of purchase and launch. They also bear the burden of modifying the bare interior of a new habitat to hold the machinery vital to their industry. Thus, the good tenant needs financial help.

Funding the Spinship: Intermediary Investor Supports the Tenants

Intermediary investors are brokers that help fund the tenants. Some end users need only a small part of a habitat; for example, a processor tenant may deal with only a portion of the ore, and use only one floor of a habitat. A farmer may occupy half the levels on a number of habitats. The broker acts to reserve habitat space on the spinship before it is built. The broker can buy or lease several habitats, and contract with tenants to modify the habitats to fit tenant needs. Instead of requiring a tenant to buy a habitat, or portion of it outright, the broker can accept payment over time, like a mortgage on a home. This system permits an expert tenant with insufficient capital to become a vital member of the spinship community.

Details of How Each Entity Makes Money

There are several ways to imagine the payment system for ROI and favorable operation for all concerned.

1. The simplest ROI system is where the tenants pay the spinship owner. In this scenario, there are no middlemen. This is unfortunate, because the vast majority of tenants can not afford the cost of habitat purchase and habitat launch, added to rent.

2. A variation on the simplest payment system is where governments pay for habitat purchase and launch, and tenants pay rent for the portion of the habitat they occupy. Yet, how can a farmer tenant afford rent? Very well, it turns out, because a farmer in a space habitat competes with costly launches from Earth of food, water, and air. The farmer can undercut that competition easily. However, the farmer is not

getting rich within the economy of space, because the farmer must pay the refinery for soil, organics and water. The refinery costs are not borne entirely by the farmer, because the refinery sells its other products, such as platinum, gold, and other rare minerals to Earth- based buyers. The refinery also sells to spinship-based makers of products ranging from electronics, batteries, solar cells, and space ferrycraft. The refinery sells fuel to power the ferrycraft, mining vessels, and barges of the asteroid ore recovery trades.

3. The third way of imagining the money transfer system is to incorporate the concept of broad freedom to enterprise. In this economic framework, there are all sorts of players. We have the easily defined owners of the spinship who collect rent from either tenant or middleman. We have the tenants. Defining the tenant group is facile. Tenants are the workers who produce the goods and services which any and all economies require. Tenants, as a group, have customers on the spinship. They have sales to other orbits, such as current satellites needing fuel, electricity, or removal to graveyard orbits. Tenants have customers on Earth. The third group of players, the middlemen, are more numerous, crafty, and capable, than a government give-away of habitats.

The middlemen are the intermediary investors, brokers, who currently inhabit the halls of Wall Street, banks, venture capital firms, real estate investment entities, and anyone with a few hundred million who wants to invest in a growing, and ever-growing, market. I call these people brokers, because they broker deals between owners of the spinship, tenants, and those outside the spinship.

While there is no typical broker group, we can see how a broker succeeds in helping others. A broker will invite various tenants to a habitat. A savvy broker will seek a refiner of asteroid ore, and a farmer, and the engineering outfit that can make a seamless system that takes ore, refines it, and delivers portions of it to the farmer, and to other consumers. The broker would seek a wholesaler of refined ore. The wholesaler would identify markets on Earth and in orbit, then take the excess refinement from the ore refiner, and distribute it to customers. One of these customers would be a manufacturer in the same habitat, who makes parts for the expanding production of mining ships heading to the LaGrange points in Earth's orbit around the sun.

Investment Summary

The renters who own a habitat, or two, pay the spinship owners for a place in space with gravity feel. Renters who do not own a habitat pay rent or mortgage to the intermediary investor who does own the habitat. The renters derive income from the services and products they provide to other tenants, to earth inhabitants and industries, and to other space-based customers. All tenants will rely upon successful asteroid mining, either as owners in mining endeavors, as suppliers, or as user-buyers of mined and processed material.

Section E: Miscellaneous

Chapter 9
Space Dev Timeline

Timeline

1. The First Age of Space: Government Projects

This age was marked by government hegemony. It takes us from Chinese rockets, around 900 or a thousand years ago, to the year 2000, when the commercial age of space might be said to have finally been born.

We saw the first glimmer of commercial space, when the first private sector communication satellites were launched in the 1960s and 1970s. This contrasted with the overwhelming focus on the Apollo program, NASA, the Russian space launches, and finally the ESA (European Space Agency), from the late 1950's until 2000. The heritage of the pork-barrel side of the Apollo program is still with us in the form of the Space Launch System (SLS) project. Up until about the year 2000, the government tightly controlled the age of space. Yet, great strides were made, such as the moon landings, Skylab, Mir, probes to Mars, and the dream of an ISS. Today, we look forward to the rise of private sector space development. The private sector is deep and broad, and touches all of us. This sector will find propitious endeavors in the Second Age of Space.

2. The Second Age of Space: Rise of the Private Sector

Our current age of space started around the turn of the 21st century, and has been characterized by the advent of private sector initiatives, notably those of SpaceX, Bigelow Aerospace, Blue Origin, the Rutan boys and Paul Allen, Sir Richard Branson, the Ansari X-prizes (starting in 1996, as the X Prize), Planetary Resources, Deep Space Industries, and many dark horses, like Rocket Lab, the space elevator project, the Skylon project, and JP Aerospace.

During this age, we should see:

a) NASA cutting edge exploratory projects, notably on the moon; the completion of the SLS project with its Orion capsule; development near or on the moon;

possibly improvements to the ISS, if not by government, then by private hands. We may also see the Russians build their own space station, and hopefully see the first commercial space station, perhaps by Bigelow Aerospace, or, Blue Origin, SpaceX, Axiom, or others.

b) Other efforts include reaching for asteroid material: Japan's JAXA Hayabusa 2 (2014-2020); ESA's Marco Polo-R, asteroid deflection project. It is slated to arrive at an asteroid in 2022; NASA's OSIRIS REx (2016), if fully funded.

c) We should soon see a space cleanup project that removes debris from earth orbits. This should be an international effort, with cleaning bills proportioned by debris contribution by country.

d) Hopefully soon, Planetary Resources, or Deep Space Industries, or others taking their place, will identify and recover the first commercially valuable water, organic, or metallic material from a near-earth asteroid.

e) We can look forward to construction of the first rotating real estate in orbit, preferably a private sector spinship.

f) Expect berthing on a spinship of self-sufficient habitats.

g) A healthy mining trade is coming, one in which asteroid wealth will protect earth from unsafe objects in space.

h) The advent of very low cost access to orbit is on its way. My bet is on JP Aerospace, but SpaceX, Blue

Origin, and the Skylon project are among the potentially viable contenders.

i) We can reasonably expect a robust aerospace industry that leads the world to a richer future for all people.

The path that gets us from the current age to the third age of space, where spinships will be in many orbits, comes with a complicated dance of space mining, spaceship development, and ferrycraft workhorses.

The first spinships will locate in LEO, and according to Al Globus and Tom Marotta, the orbit should be in equatorial LEO, or ELEO. This makes practical sense, because a new spinship will rely upon existing rockets to receive the initial supplies of personnel, food, refinery and fabrication machinery. The spinship orbit is somewhat near the ISS orbit, but in such different planes that delivery to one space station can't easily be shared with the other.

With a spinship in ELEO, asteroid ore barges will find far less efficient delivery than in a mature system of ferrycraft and translunar hand-off orbits.

The first spinship can be built to a smaller size than is optimal for earth gravity and dizziness-free use. Al Globus of NASA-Ames, and Tom Marotta have a book, "The High Frontier: An Easier Way," which explains such an optimal system. Later, the first ship can be moved to translunar orbit to become the major refinery center for asteroid resource recovery.

3. The Third Age of Space: Local Dispersal

Sometime after 2050, the human race will enter an age populated by spinships in far-flung locations. While many ships will occupy productive earth orbits, many more will orbit the moon, Venus, Mars, and occupy the asteroid belt. Intrepid exploratory ships will test survival and technical limits beyond the belt.

This exploratory age will provide many a great opportunity for discovery, trade, science, and technology. Yet, if we venture beyond the belt, we find insufficient sunlight to power our systems. To travel beyond, we'll need energy extraction means not yet commercially available. There are some candidate energy sources. Notable are the NASA projects in cold fusion.

4. The Fourth Age of Space: The Great Dispersal Beyond 2090

In this age of space, whose beginnings one can only guess at, spinships will venture far beyond the asteroid belt. Some ships will take orbit around Jupiter and its moons, Europa, Ganymede, and others. A few ships may arrive at distant Saturn. These ships will need local sources of energy, because traders from the asteroid belt will not be inclined to make regular visits to the far reaches of Saturn. A great source of fuel may be found in the methane rich mantle of Titan, Saturn's largest moon. This age will realize fully, that, even if you can recycle everything onboard the ship, you still need energy to do so, and to fuel the ship.

The approach to both Jupiter and Saturn would probably not include a visit to Jupiter's Trojan asteroids, nor to Saturn's comparable asteroids. These space rocks are found at LaGrange points ahead and behind Jupiter and Saturn in their orbits around the sun. Unfortunately, getting from Jupiter's Trojans to Jupiter is a longer voyage than going from Earth to Jupiter. It is more likely that resources will be derived from moons of Jupiter and Saturn.

Sometime in the 22nd century, a few highly advanced ships might go farther out than Saturn, and finally reach the odd-tilting Uranus, or even Neptune, if they have a source of energy and raw materials. Life-sustaining organic and other material might come from the moons that surround the gas giants, or from the planets themselves.

During this long age of great dispersal, reliance upon supply ships from the inner solar system will surely end, only because speed of travel will reach a kind of plateau. While there may be occasional trading ships traveling to a specific concentration of outer planet spinships, such traders as may come will probably be explorers.

The tail end of this age of space is grasping the hand of the next, and who can tell where one ends and the other begins.

5. The Fifth Age of Space: Spacekind

When the tribes of proto humankind roamed the earth, genetic isolation and genetic recombination ruled alternately. When tribes were few, and intermingling was of low frequency, humans developed a multitude of varietal characteristics. From small isolated populations came promotion of DNA groups, not just of types of hair or skin color, but of genetic tendencies, such as diabetes, malaria, and recognizable facial and stature differences. These characteristics were so distinct, that anyone might say that this particular person or that one belonged to a named tribal group.

By contrast, our own age, the 21st century, may be a time of great mixing. In this century, we may all tend toward looking alike. Skin color will move toward a nice tan.

We are all very closely genetically related, differing mostly in skin color, which occurred, some authorities claim, not all that long ago. In just a few centuries, if we remain only on earth, we may come to the end of our DNA differences, and then our species will decline rapidly, all of us being so closely related that our offspring will be less robust. If one becomes sick, all may sicken. We may become a mono DNA species, a mono-bio-culture. How will we survive? How will we declare: Vive' la difference.

The answer may be found in the natural mechanism of the fifth age of space.

In the fifth age, through the process of dispersal, we will once again live in small, isolated communities. This isolation is not complete. We will have infrequent visitors from other spinships. Yet, the isolation will be the very thing that saves us, strengthens us as a species.

During this age of great tribal separation, ships will disperse to all distant corners of the solar system. Each ship becomes a community of DNA, with its own characteristics, including bio-processes of novel DNA recombination. Each ship community will express some DNA set that may be significantly different from the DNA of other spinship communities. This will give each space-based tribe such traits that are indescribable at this time. Possibly some computer modeling of DNA changes, or an inventive science fiction writer may guess at the DNA descriptions. It is by the natural mechanism of distant neighborhoods, that the human race will acquire greater breadth and depth of DNA than it currently possesses. The occasional visiting ship will allow a mixing of DNA that will add to the strength of the human race, such as a hybrid is stronger than either parent. And, we are all hybrids. So far.

As humanity returns to its roots of greater isolation, the alternating of small and large group mixings renews the human race. We become a product of living in space. Humankind becomes spacekind.

6. The Sixth Age of Space: Metakind

Beyond the fifth age of space, progress in extending lifetimes will skew our communities toward the mature person of 150 years of age, or older. Tinkering with DNA, or the mere recombining of so many different people, may produce a human being whose DNA allows for each cell to divide less often, or for many hundreds of divisions, extending the life of the cell, and thus, the life of the individual. Various choices of DNA may be available to the consumer looking to be improved. What these changes might be are explored in sci-fi and fantasy tales.

Glossary

Ascender: One of several V-shaped wings filled with a buoyant gas, usually helium. One type of ascender travels from the ground to the dark sky station (DSS). The orbital ascender has the widest wingspan, and travels from the DSS to orbit, returning to a DSS almost anywhere in the world. Ascenders built in orbit can carry cargo from orbit to land almost anywhere on the earth.

Ballistic Descent: a type of re-entry to earth's atmosphere by means of falling. Ballistic descents do not involve the use of rockets to slow descent to the ground.

Dark Sky Station: The DSS is a floating station at the edge of space. Its altitude varies from 100,000 to 140,000 feet above mean sea level. The DSS arms are filled with a buoyant gas. The station looks like a starfish with five cylindrical arms meeting at a flat central platform. Its purpose is to invite tourists to enjoy the view and stay a few days in luxury suites. The DSS can also act as a boarding station, when airships to orbit dock along one of its five arms.

Helium: An inert gas used to make high altitude ascenders buoyant.

Ion Engine: A space-worthy engine that uses an inert gas expelled as ions at high velocity to propel it to various points in earth orbit, or beyond. An engine that powers mining craft in the mining trade between earth and the asteroids.

ISS: The International Space Station. It occupies a tight range of LEO orbits, averaging about 240 miles up (nearly 390 km).

LEO: Low Earth Orbit. A range of orbits starting at 160 km (approx. 100 miles) above the earth, and extending to 2,000 km (approx. 1,243 miles).

Space Ferry: A craft that stays in space and shuttles cargo and passengers between very low earth orbit and the spinships in higher LEO. The ferry also heads out to spinships in translunar orbit. The ferrycraft is also called, an OTV, Orbital Transfer Vehicle; also called, a space tug. Ferry craft serve the asteroid mining trade by accepting ore, water, and organics from mining barges in translunar orbit, and taking the cargo to spinship, to the ISS, or to LEO for delivery to earth markets.

Spinship: A station in space that rotates. A community of habitats inside the perimeter of a disk in orbit. A rotating, self-sufficient community in space. A spaceship that rotates to provide the feel of gravity.

The Space Trade: Business conducted in space. Any business in orbit. Examples: business on or between spinships; the ferrycraft business; the asteroid mining business; space tourism; the satellite moving business; the launch to orbit business; the business of delivery of goods to the Earth. The system of exchange of goods and services in, from, or to orbit. The system of economic relationships of space and Earth-based businesses.

Translunar Orbit: Referring to traveling around the earth in an orbit that is bigger than the orbit of the moon.

Addendum

Choice of Law

The following is general legal information, not meant as legal advice.

Here we come to a gnarly issue. Who are we when we become spacekind? Are we so out on the frontier that frontier justice might obtain? When we occupy Earth orbit, it is doubtful that we would welcome the often lawless nature of the old west. Yet, that's what we will have, and worse, if we don't solve a certain troubling issue. Namely, we must find an operational legal process tailored to space living, and the commercial needs of a spinship city at the edge of the cosmos.

When a problem occurs, it's good to have a framework to solve it. This holds true in fashioning a new society. So let's start with a framework we know works.

Our country began, not with an empty slate, but with a rich political palette from which to choose a form and substance of government. We adopted much from our British heritage, in terms of institutions of government, especially the ideas of a legislature, prime minister, and king, or, in our case, not a king but a president. For existing law, we adopted the English common law, where each case was decided on the basis of custom, and the cases before it, and as modified by any standing law. Such standing law might have been king-made, or perhaps a law passed by one of the many parliaments over the centuries since 1137.

Our legal framework was supported by a number of philosophies. From France came the philosophy of Montesquieu, and others. From England came John Locke, who insisted that the social contract between government and the governed is foundational; the government only has power that the people give it. Locke also stood for the freedom to travel from one country to another, showing that people could vote with their feet. The point is that we started with a framework. We should start the community of space and the spinship with a framework.

To suggest one framework would be to deny other worthy ones. However, because we must start somewhere, let's propose the adoption of the laws of a certain state that are both difficult and wonderful to grapple with, the body of law of the State of California. In fact, it might be wise to request annexation by the State of California, and seek a city charter, as part and parcel of the new found land. In this way, the spinship might have all the advantages that communities have at their beginning in that State.

It must be added that there are some California laws which should not be applied to the spinship, but that is the subject of an entire book, and it will have to wait.

Going out to left field, a spinship could accept the principles of the state of Jefferson. This is a mental state, existing primarily in far northern California, southern Oregon, and, no doubt, in other places around the nation and the globe. Jeffersonians love freedom of the individual. My best vote is to make the first spinship a new State in the Union, the State of Jefferson.

For those companies who pioneer these new space ventures, it is suggested that they choose carefully their corporate form. After some searching, the best and most freedom-giving of the various corporate forms is not the benefit corporate form, but the flexible purpose corporate form of California. This form gives the most leeway and flexibility of any law of incorporation extant as of late 2013. To the author's knowledge there is not another law called by such a name, nor one that contains such friendly language. A few states have benefit corporation forms that approach the freedom of choice that California's flexible purpose law gives, but in no state is there a law equal to California's. This is, of course, a matter for each company's lawyers to help decide, and it is left to their competent minds to investigate.

The corporate form can support finance of the space trade, by specifying that quarterly profits are not the most important goal of the corporation.

Having suggested a framework, it might be wholly expected to be modified as events unfold. We must remember that our human-made laws are meant to help us in challenges political, social and physical. In space, the same is true. We are just not able to predict all the ramifications of space-based living. This means that we will operate in a somewhat piecemeal fashion, as space denizens move to discover new laws that are practical and tailored to space life. We know that the natural laws of space are hard and unchangeable, and we must meet that environment with courage, patience, and understanding. We can do that. We will try to take each step as it comes, and look forward to each tomorrow with the knowledge gained from the day before.

CAVEAT ITAR

What do you tell your would-be investors around the world about ITAR? Tell them to get an attorney competent in ITAR restrictions. ITAR is short for International Trade and Arms Restrictions. It is a law that fetters free exchange of technical information among countries and individuals. It is both domestic and international law and treaty. It prohibits you from talking technical to foreigners. Waivers to share tech data are given in many fields, except space technology. That is forbidden. General educational knowledge can be shared, but when specific technical details are shared, then you're in trouble. A number of highly placed corporate officers have served prison time for seemingly innocuous presentations to a group of foreigners. The U.S. State Department administers the restrictions.

ITAR is meant to protect American security, but I hope someday that Congress changes the law to reflect a warning by Benjamin Franklin, who said, "Those, who trade freedom for security, deserve neither." And, while on that note, let's remember, every day, to provoke the universe, and stand on the American foundation given us by President Thomas Jefferson, who said, "Dissent is the highest form of patriotism."

The Next Book of The Space Trade Series

The next book, "Spaceships of the Space Trade," will reveal necessary functions, design requirements, and general construction suggestions for the five essential types of space vessels in space trade commerce.

About the Author

L. Paul Turner flew jets for the US Air Force, studied law, real estate planning, and legal history, earning his doctorate. During law school he was Director of Environmental Affairs for a local Coors wholesaler, and managed one of the first recycling centers.

After teaching law and economics, and winning his first appellate case, he pioneered the first webinar designed especially for California State Bar approved credits. He is the author of two construction law books. After serving as Educational Director for a local bar association, Paul returned to recycling to collect and analyze data for California's unique and successful bottles and cans program.

In 2004, Paul teamed with JP Aerospace to build experimental airships and propulsion systems for high altitude flight. Paul can be contacted at The-Space-Trade-Update.blog.

Dedication

This book is dedicated to Dr. Gerard K. O'Neill, who wrote the great dream book, "The High Frontier." "The Space Trade Update" describes some of the lesser dreams that Dr. O'Neill seemed to suggest as foundational to the great dream.

Today, the billionaire space developer is the stand-out leader of the space trade. While this may always be so, standing in the wings is the broader investment community, which can include ordinary people, like the neighbor down the street.

We stand only at the start of the commercial space age. As this age matures, we, the people, will be the ones who build, own, and operate spinships, to the benefit of ourselves and the entire human species. Therefore, I dedicate this book to all of us.

Inspirations

Elon Musk of SpaceX, Al Globus of NASA-Ames, Keith Lofstrom, Robert Bigelow, the National Space Society, the AIAA, Engineer Adam Koppy, all have inspired me in the making of this book. I would especially like to note a leader in aerospace, an independent thinker whose unique method of achieving orbit may make the age of launching rockets entirely obsolete. This is someone who has sent more free student experiments to the edge of space than anyone, even NASA. This man is John Powell of JP Aerospace. It was he, who initially inspired this series of books. Explore, when you have time, the arcane knowledge applied to aerospace at www.JPAerospace.com. Thank you, John Powell, for your pioneering work. You are succeeding in an industry where life and work take many stumbles before we get anywhere.

Thank You

I wish to thank my many readers and listeners, who demonstrate every day that the human race cares about its future, and its continued existence in the cosmos.

And, I thank my friends, who helped to make this book a more intelligent and concise presentation on the path to space than was the first tome. Especially, I wish to thank Adrienne Ellis, both in her editorial insistence on concise writing, and in her wonderful, and easy to listen to voice, as narrator. Without her help, this book and audio book would not have been possible. Thanks go to Robert for his expert technical assistance and advice. Thanks go to my family for putting up with my writing efforts. Thanks to Matt for his requirement of concise writing. I still struggle to attain the high bar he sets. Thanks to my colleagues at Sacramento L-5 Society for their continued insistence that we occupy space, and their provoking comments on all things scientific and science fictional.

Final Note

The audio book "The Space Trade Update Audio" is available for your listening pleasure on Audible.com. Noted screen and voice actress Adrienne Ellis was kind enough to speak as narrator for the audiobook. Thank you, Adrienne for your great and pleasant voice, and continued help in education and entertainment.

- L Paul Turner